GLEN ROCK PUBLIC LIBRARY
GLEN ROCK, N.J. 07452

INSECT MIGRATION

All animals migrate. A migration is any planned journey from one place to another. This book describes some interesting and extraordinary migrations ranging from a few inches to 2,000 miles.

MIGRATIONS

INSECT MIGRATION

Liz Oram
and
R. Robin Baker
Department of Environmental Biology
University of Manchester

STECK-VAUGHN
L I B R A R Y
A Division of Steck-Vaughn Company
Austin, Texas

© Copyright 1992, this edition, Steck-Vaughn Co.

All rights reserved. No reproduction, copy, or transmission of this publication may be made without written permission from the publisher.

Library of Congress Cataloging-in-Publication Data

Oram, Liz, 1964–
Insect migration / Liz Oram and R. Robin Baker.
  p. cm. — (Migrations)
Includes index.
Summary: Discusses how insects such as wasps, army ants, locusts, and various butterflies and moths migrate and the purpose of these voyages.
ISBN 0-8114-2926-1
1. Insects—Migration—juvenile literature. [1. Insects—Migration. 2. Insects—Habits and behavior.] I. Baker, Robin, 1944– . II. Title. III. Series: Oram, Liz, 1964– Migrations.
QL496.2.072 1991   91-12776
595.7'0525—dc20   CIP AC

Cover: *A butterfly tree in Mexico covered with monarch butterflies.*

Typeset by Multifacit Graphics, Keyport, NJ
Printed in Hong Kong
Bound in the United States by Lake Book, Melrose Park, IL

1 2 3 4 5 6 7 8 9 0 HK 96 95 94 93 92

Contents

Introduction 6
  Ways of Migrating

1. Insects with a Home 9
  The Caterpillar
  The Honeybee
  The Bee Dance
  The Female Hunting Wasp

2. Moving to a New Home 13
  The Caterpillar
  Army Ants

3. Blown by the Wind 17
  Aphids
  Leafhoppers
  Moth Caterpillars

4. Butterflies and Dragonflies 21
  A Flying Visit
  The Small White Butterfly
  The Painted Lady Butterfly
  The Monarch Butterfly
  The Wanderers
  Dragonflies
  Finding the Way

5. Moths 30
  Moths and Lights
  The Large Yellow Underwing Moth
  The Bogong Moth
  The Silver Y Moth
  How Moths Find Their Way

6. Locusts 36
  Hoppers
  Solitary Locusts
  Swarming Locusts
  Locust Devastation

Glossary 43
More Books to Read 44
Picture Sources 44
Index 45

Introduction

Have you ever noticed tiny creatures flying and crawling around, especially during warm weather? Maybe you've seen brightly-colored butterflies fluttering around, or tiny orange ladybugs scurrying along in the grass. You've probably also seen and heard bees and wasps buzzing around nearby while having a picnic in your yard.

To you, these creatures might all have seemed very different. Yet in

How do you recognize an insect? If a creature does not have three pairs of legs it is not an insect! Most insects also have two pairs of wings, and a body divided into three parts.

With the coming of fall the monarch butterflies have returned to the California town of Pacific Grove. Every year the town celebrates their arrival with a parade.

one way they were very similar. They were all *insects*.

What is an insect? Insects are creatures whose bodies are divided into three almost separate parts—head, thorax, and abdomen. The skeleton is on the *outside* of the body. They have three pairs of legs, and usually two pairs of wings.

Not all "creepy-crawlies" are insects. Spiders, for example, are very different from the creatures described above.

Insects go through a series of amazing changes during their lives. When they begin life, their form is completely different from that of their parents. The change, called "metamorphosis," usually has four stages. First, the female adult lays

eggs. An egg hatches into a larva, which can move around and feed. Then the larva changes into a pupa, which cannot move or feed, but extraordinary changes take place inside it. Finally, the pupa splits open and a young adult comes out. It flies away to begin a new life.

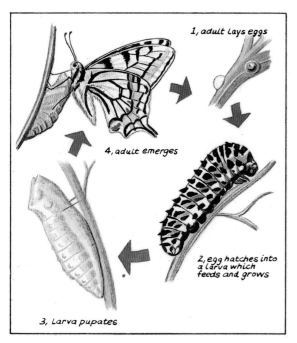

A butterfly lays eggs on a plant. Each egg will hatch into a caterpillar. The caterpillar turns into a chrysalis. From the chrysalis a butterfly emerges. Soon it will lay eggs . . .

Often the larva stage lasts a lot longer than the adult stage. For example, a dragonfly larva spends about two years at the bottom of a pond, but the winged adult lives only about two months.

Did you ever wonder what insects are doing as they fly and crawl around? Almost every insect you see or hear is making a journey. Some may be on their way to or from their homes. These insects may travel only a few yards and are probably collecting food. However, not all insects have homes. Some spend much of their time on the move and make much longer journeys. There is a very famous butterfly, called the monarch, that travels thousands of miles every year. You can read about this in Chapter 4.

Ways of Migrating

Insects that make journeys are called "migrants." The journeys they make are called "migrations." The longest migrations are made by insects that can fly. However, some insects migrate by walking. Some even carry their homes with them when they migrate. Needless to say, the "walkers" do not travel as far as the "flyers."

Sometimes during the day, it is easy to watch insects migrate. Some insects such as moths, however, migrate only at night and are difficult to see. Some insects travel completely alone; others migrate in large groups. One type of insect, called the locust, migrates in enormous "swarms," or groups. In such swarms there may be millions of locusts.

In this book you will not only find out about the migrations of the monarch butterfly and the locust, but of many other different kinds of insects as well.

1 Insects with a Home

Most of you have seen caterpillars in your yard or the park. Caterpillars are young insects that nearly always have some favorite place where they spend most of their time. This place is a kind of home. Sometimes their home is just some favorite position out in the open on a leaf or twig. Some caterpillars actually build a kind of shelter around themselves, as

Some caterpillars spin webs, like spiders do. Sometimes a group of them will work together. They will use this web as a shared home.

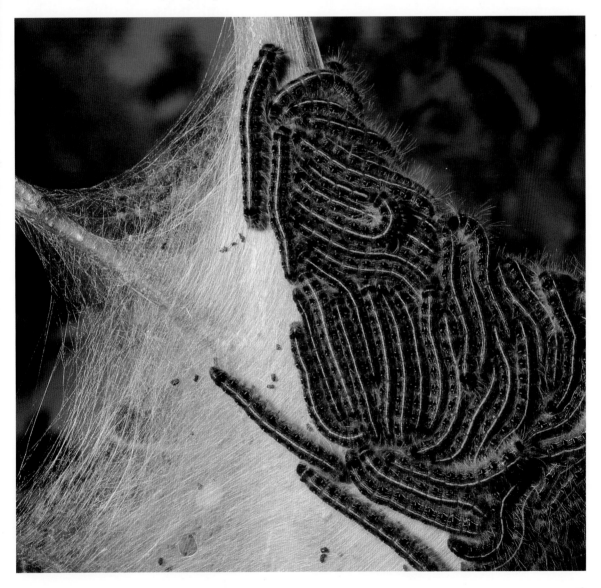

you will see later in this chapter. Whatever their home is like, it is where they spend most of their time.

Caterpillars are not the only insects with a home. Some insects live alone in their home; others share it. In this chapter we shall be looking at caterpillars, bees, and wasps. The journeys they make are short and not nearly as spectacular as some of the journeys we shall read about later. Nevertheless, they are still migrations.

The Caterpillar

Have you ever wondered what a caterpillar's home looks like? You may be surprised to know that some caterpillars, like spiders, spin webs. Probably some of the webs you have seen have belonged to caterpillars. Although spiders will spin webs almost anywhere, caterpillars almost always spin their webs among leaves. Sometimes a group of caterpillars that have hatched from a cluster of eggs on the same leaf will spin a web together. This kind of home is called a "tent" home.

It is possible to find bushes covered with caterpillar webs. This is an impressive sight, especially in the early morning when the webs shine with dew. Caterpillars only really need to leave their homes to find food. Because they feed on the surrounding leaves, they do not have to go very far for a meal.

The Honeybee

Honeybees live in very large groups called colonies. The place where they live is called a "hive." You may have seen a hive yourself. However, it is more than likely that the hive you saw was made by humans. These hives are very different from those that are made by honeybees in the wild.

In the wild, bees often build their hives in holes in trees. A large colony may contain thousands of bees. They all live together in the hive. Unlike wasps, which are carnivores, bees are vegetarians. They live on a sugary liquid called "nectar." Nectar is collected from flowers. The bee does not eat all the nectar it collects right away. Instead, the nectar is carried in the bee's mouth back to the home hive and put into special "storage compartments." Eventually, some of this stored nectar is fed to the developing larvae, and some is eaten by the other bees in the hive themselves.

The Bee Dance

Most of the journeys made by honeybees are to find and carry nectar back to the hive. Bees have developed a spectacular way of telling each other about the discovery of a new food source. It is like a kind of insect language that we humans are only just beginning to observe and understand.

The "discoverer" flies home and performs what is known as a bee "dance" for the other members of the hive. There are two dances. The first is called the "round" dance. It tells the other bees that a new food source has been found close by. The other dance, called the "waggle" dance, is performed when the newly discovered food is much farther away. The more "waggles" the dance contains, the farther away the food is to be found.

Honey is the only human food made by insects. In the past people would raid the nests of wild bees. Now beehives make collection much easier.

The Female Hunting Wasp

Some wasps live on their own. Others live in large groups called "colonies." The female hunting wasp nearly always lives on her own. Her home is a tiny burrow that she digs in the ground. This burrow is used as a kind of storage house. As

A female hunting wasp spends most of her life collecting food for grubs she will never see. When her eggs hatch into grubs they will find this spider waiting as their first meal.

soon as the burrow has been dug, the wasp sets out to fill it almost to the top with food. Then, right on top of this pile of food, she lays her eggs. Perhaps "wasp pantry" is a more suitable name for such burrows.

Wasps are carnivores. This means that they eat other animals—usually other insects. Once the wasp has caught her prey, she injects it with a special liquid from her sting. This liquid paralyzes the insect but does not kill it. The wasp then carries the insect home and stuffs it into her burrow. Only when the burrow is almost full does she lay her eggs. When the eggs hatch, the grubs feed off the paralyzed insects.

Some wasps have more than one burrow, and lay lots of eggs. These wasps are kept quite busy flying around looking for food to put into the different burrows. With all this flying around, it is amazing that the wasp hardly ever gets lost. How does the wasp avoid getting lost herself, or losing her burrow?

The hunting wasp is a very clever insect. After digging her burrow she may spend some time exploring the area around it. By doing this, the wasp gets to know her home territory very well and is able to recognize her burrow easily. She may notice certain things in particular, like a clump of grass or a collection of stones. When she goes on a hunting trip the wasp uses these things as "markers" to help guide her back. The wasp also selects a marker for each individual burrow. Just before she sets off on each hunting trip she may fly around once more to make sure she knows exactly how to find her way home and how to recognize each burrow.

Sometimes, especially if food is scarce, the wasp flies farther away from home than usual. If this happens, she uses the sun to help guide her, until she is close enough to home to spot her own markers.

2 Moving to a New Home

In the last chapter, we read about some insects with a home. We also read about their tiny migrations as they went out from their homes to eat, and then returned. In this chapter we shall find out that, every so often, insects move to a new home. They leave the place where they have been living and go to live somewhere else.

Most insects, like the caterpillar, do not carry anything with them when they move. Army and driver ants, however, are very special types of insects that do take things with them when they move. First we shall look again at the caterpillar and how it changes its home.

Caterpillars that live in groups will often migrate together. They travel in a procession until they reach a new source of food.

13

The Caterpillar

As we have seen in the previous chapter, caterpillars have a permanent home most of the time. However, once they have eaten all the leaves that surround their home, they need to move to a new place. This new home has to be where there is a good supply of leaves for them to eat. Caterpillars that live together, and that have spun a web together, also move to a new home together when the time comes. They move in a long procession until they find a new leaf or twig with lots of uneaten leaves surrounding it. The caterpillars will remain in this home until they have eaten all the surrounding leaves, and are forced to move again.

Army Ants

Army ants are both fascinating and frightening. They live in parts of Africa and South America. Sometimes they are called driver ants. Army and driver ants are bigger than most other types of ants. They also move faster over the ground and travel farther when they migrate. In addition to this, army and driver ants do something very special that other ants do not do: they move to a new home every night.

Just like other ants, army and driver ants live in colonies and build nests. However, even though they

Army ants migrate almost every night, carrying their grubs with them. They will eat any creature that doesn't get out of their way.

Nothing stops a march of army ants. Here a column is crossing a stream by clambering over each other's entangled bodies.

are larger than most other ants, they build much smaller nests. Indeed, they may be no more than a pile of leaves and twigs and are nearly always built on the ground. Inside a nest, the grubs, the queen ant, and her "workers" can be found. The food that has been collected by the worker ants is also there.

Each colony has only one queen ant. The queen is the largest ant in the colony. She is about three times the size of all the other ants, and spends most of her life being waited on by the workers. Besides the queen ant and the workers, the colony also contains "soldier" ants. Soldier ants have very large and powerful front claws, and their main job is to protect the colony.

It may seem strange to you, but all the ants in a colony are related. This is because they all have the same mother—the queen ant. All the eggs in an ant nest are laid by the queen. These eggs eventually hatch into grubs, which then grow into ants.

For most of the time, the colony is made up of only female ants, all of whom are sisters. The queen produces male offspring only at certain times. This is usually a very short period during a rainy season.

Nearly all the journeys made by ants during the day are to gather food. This job is done by the worker ants, the busiest ants in the colony.

When an ant discovers food, it leaves a smell on the ground leading from the food back to the nest. The worker ant makes this "scent trail" by dragging the tip of its abdomen over the surface of the ground as it migrates back to the nest. The other members of the colony are able to find their way to and from the new food source by following this scent trail. Each new worker using the trail also leaves scent on the ground. This makes the scent even stronger and,

Ants will stop marching while the queen lays eggs. Meanwhile they forage for food. Here a wasp larva is being dragged to the nest.

therefore, even easier to follow. When the food at the end of a trail runs out, the workers stop leaving scent and the trail soon disappears.

Great amounts of food are brought back to the nest by the worker ants. A very large nest may have five or six scent trails leading from it at a time.

However, there is rarely enough food in the surrounding area to feed the ant colony for longer than one day. So, as soon as night falls, the whole colony migrates. It travels in a long, army-like procession, taking the nest of grubs with it. The grubs, which are unable to walk, are carried by the worker ants in their mouths. It is certainly a big operation. The trail of moving ants may contain as many as half a million insects.

In one night the colony may travel as far as a hundred yards. This may not seem very far to you, but just think how far it must seem to a tiny ant. As the procession of ants marches along in the darkness it is almost unstoppable. Anything that stands in the way is likely to be eaten alive—not only other small insects, but any larger animal that is unable to get out of the way.

The ants continue until they find a suitable place in which to build another nest. This is home only until the following night, when migration starts again.

3 Blown by the Wind

In this chapter we shall be looking at the migrations of very small insects. Aphids, leafhoppers, and the very small, hairy types of moth caterpillars are found in nearly all parts of the world. They are so light that when in the air these insects are blown along by the wind and do not need to flap their wings. When they move to a new home they just allow themselves to be blown from one place to the next.

On hot days, sometimes these insects, especially aphids, land on people. You may have noticed them on your skin or clothes. They look like little, colored specks and are very easily brushed away.

Aphids

Aphids are very tiny insects that are usually green or black in color. If you look carefully at any roses in your garden you might be able to see hundreds of them clustered on a single plant.

Aphids feed off a substance called "sap." Sap is found in the leaves and stems of most plants. The aphid's mouth is specially designed to suck sap out of the plant. This special mouth is shaped like a straw with a sharply pointed end. The aphid uses the point to pierce a leaf or stem and then sucks up the sap through the straw. If a plant is attacked by a lot of aphids at the same time, it usually

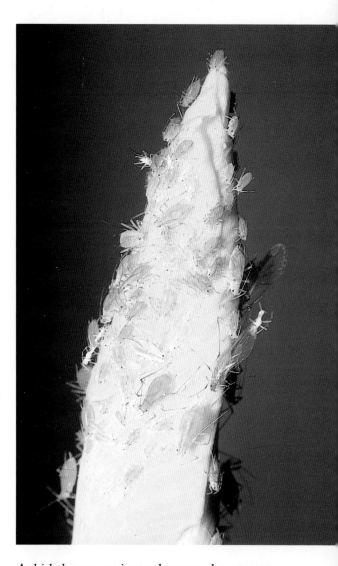

Aphids have to migrate because they use up their food supply by breeding so quickly. A single female can produce 25 daughters a day. These can begin breeding in eight days.

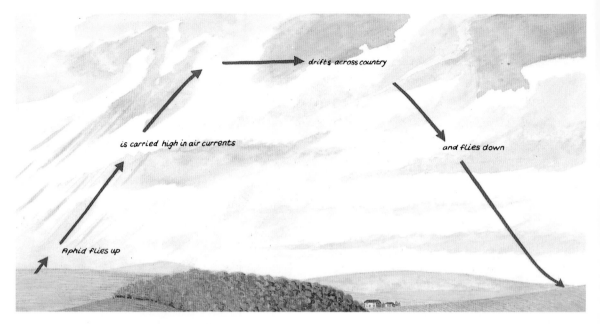

dies. For this reason, aphids are disliked by most gardeners and farmers.

Aphids live for only a few days and most, though not all, migrate only once during their lives. They might wish to move to a new home because the plant on which they are living has become overcrowded. The aphid starts its migration by flying up into the air. However, it flies only until it is just above the level of any surrounding trees. After this, it stops flapping its wings and lets the wind blow it along.

The wind carries the insect along above the trees. At this height the aphid is safe from being blown into spider webs. However, it is not completely safe. There is a good chance that it will be eaten by a swallow. Swallows catch and eat insects on the wing.

When the aphid sees a suitable new home on the ground below, it

Only the winged, female aphids migrate. They depart in two main flights, one in the morning and one in the afternoon, but only if the temperature is above 63°F.

begins to flap its wings. It flies downward out of the wind and on to the greenery. After landing, it selects a new plant as its home.

You may be wondering how far an aphid may be blown by the wind. If there is only a light wind, the aphid isn't carried very far before it flies downward. Its new home may be only a few yards from its old home. However, in very strong winds, aphids can be carried hundreds of miles.

What about the height at which the insect is carried? Occasionally, in certain kinds of wind, aphids can be carried to amazing heights. They have been found migrating at the same height as airplanes. A number of experiments have been performed to show this. In most of these experiments a large net was suspended from the underside of an aircraft, and was used to catch migrating insects in flight. However, aphids do not seem to like migrating at such great heights. If they find themselves being carried rapidly upward, most try to fly downward at the first opportunity.

Leafhoppers

Leafhoppers are found all over the world, but in this section we are going to look at a type of leafhopper found only in the United States. This is the beet leafhopper. Like aphids, the beet leafhopper also has a mouth like a pointed straw, and does a lot of damage to the plants on which it feeds.

Leafhoppers feed on the sugar beet plant. Sugar beets are grown in large fields in many parts of the country. Beet leafhoppers spend the winter in Texas in sugar beet fields where they damage lots of leaves as they hop from plant to plant. It is this hopping that has given the leafhopper its name.

In spring the beet leafhopper migrates north. It migrates the same way the aphid does. It flies up into the air, then is quickly taken by the wind and blown along. The wind carries the insect northward to other sugar beet fields. Here there are lots of fresh leaves for the leafhopper to enjoy. The insect then lays eggs and produces lots of young leafhoppers. These also feed on the leaves, and grow very quickly into adults that do more damage to the crops.

In the fall, leafhoppers migrate south again. The amazing thing

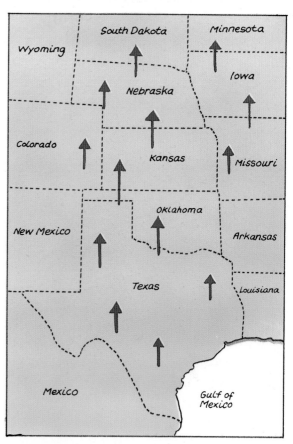

Leafhoppers can jump from one plant to another, but that is not how they migrate. They fly up into the air, then let the wind carry them along.

about these insects is the distance of their migration. They can be blown by the wind for as many as 600 miles before landing. This makes it very difficult for farmers to control these insects. Every year a number of farmers lose part of their crop to these tiny insects. A lot of money and food made from sugar beets can be lost because of them.

Moth Caterpillars

Moth caterpillars start life as eggs. They are laid on leaves, sometimes very high up in the trees. The eggs hatch into caterpillars. Some types of these caterpillars are covered with lots of tiny hairs, which makes them look like tiny balls of fluff.

If a leaf gets too overcrowded, and the caterpillars run out of leaves to eat, they migrate. To do this they simply have to drop off their leaf and let the wind carry them along to the next tree. Sometimes the wind carries the insect very high up into the air, perhaps as high as half a mile or more. The hairs on the insect allow it to be blown very easily by the wind.

Most caterpillars migrate by walking and climbing, but some drift in air currents. The delicate, silky hairs all over its body catch the air, and make it very light for its size.

4 Butterflies and Dragonflies

In this chapter we shall be looking at some insects that migrate during the daytime. Most of them migrate on their own. They are called "solitary" migrants. One such insect is the butterfly.

You may have seen brightly-colored butterflies fluttering around

Butterflies can be easy prey for birds. But when at rest this colorful comma butterfly folds its wings and looks like a dead leaf.

in gardens, especially on sunny days. Perhaps some of them were diving in and out of flower beds. Other butterflies may have been flying from leaf to leaf, inspecting one leaf at a time.

Did you wonder what they were doing? Those diving in and out of the flowers were looking for nectar (like the honeybees in Chapter 1). The butterflies inspecting leaves were probably trying to decide where to lay their eggs. Or, if it was almost evening, they may have been looking for the best leaf to spend the night under.

A Flying Visit

Most of the butterflies you see in the garden, the park, or the open countryside, are just visitors. They have stopped only to feed from the flowers, sleep, or lay eggs on the leaves. Some of the male butterflies

When a butterfly flutters from one flower to another, it may look like aimless flight but if you could follow it all day you would see that it is migrating on a direct course.

may be looking for a female with which to mate. However, whatever they are doing, most of them will leave after only a short time. This is because they are migrants, without a home. Their visit could be called a "flying visit."

When the butterflies do fly off, they are continuing a long journey, and will not return. Instead, they will fly in a straight line across the countryside until they next need to feed, rest, mate, or lay eggs. The next place they visit will be a long way away.

Butterflies are almost always on the move, migrating from one stopping place to the next. It may surprise you to know that some butterflies migrate hundreds or even thousands of miles during their lives. This is quite a feat for an insect that may live for only a few weeks. Now let's look at the migration of the small white butterfly, one of the most common types.

The Small White Butterfly

The small white butterfly can often be found where there are cabbages. This is because the female often lays her eggs on cabbage leaves. The eggs then hatch into caterpillars, each of which performs its own tiny migration.

When it is half-grown, the caterpillar crawls from the outer leaves into the heart of the cabbage. This may not sound like a very long journey. The caterpillar may crawl only 12 inches. However, even this is a long way for an insect that is less than half an inch long and can crawl only very slowly. In the cabbage heart, the growing caterpillar eats the leaves. As you can probably guess, the small white butterfly is not a welcome visitor to the gardens of vegetable growers!

The caterpillar stops eating cabbage when it has grown to full size (about an inch and a half). Then it has one more journey to make before it can become a butterfly. It

The small white caterpillar makes two migrations. First it travels from a cabbage's outer leaves to its center. Later it travels away from the cabbage, for a place to pupate.

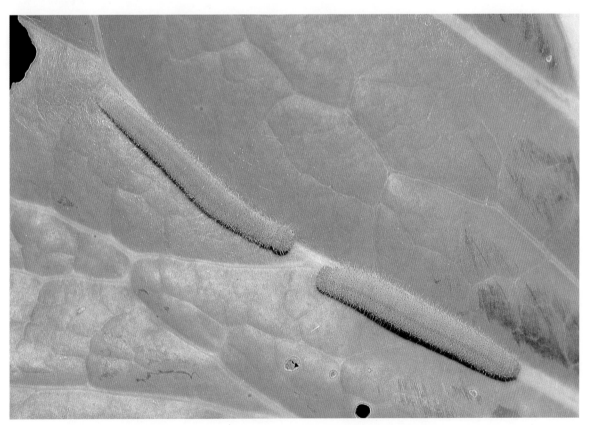

The life of the small white butterfly is shorter than its chrysalis or caterpillar. It lives for only about three weeks, but in that time it might travel 120 miles from its birthplace.

crawls down the cabbage stem onto the ground, then sets off across the soil in a straight line. The caterpillar does not stop until it finds something high, like a tree or wall, which it then climbs.

Some may crawl as far as 20 yards across the ground and still have enough energy to climb right up the wall of a house. There, under a window ledge or roof shingle, the caterpillar turns into a chrysalis. The butterfly that hatches from the chrysalis makes a much, much longer migration.

Like most butterflies, the small white spends much of its time migrating across the countryside. Only when it needs to feed and rest does it look for somewhere to land. For this butterfly, a garden with a patch of cabbages makes the perfect stopping place.

After hatching from the chrysalis, a small white butterfly lives for about three weeks. Most of that time it spends traveling in a straight line, migrating across the countryside. When it dies it may be as far as 120 miles from the cabbage it spoiled as a caterpillar.

The Painted Lady Butterfly

The painted lady butterfly is sometimes also called the thistle butterfly. This is because the female usually lays her eggs on thistle plants. The painted lady can be found in nearly all parts of the world. Unlike the small white, which lives only about three weeks, this butterfly can live as long as seven or eight months.

Since they live much longer than small whites, painted lady butterflies also migrate much farther. Some might travel as far as 1,200 miles.

In northern Africa, as soon as the painted lady emerges from its chrysalis, it begins a three-month migration.

Those butterflies that are born in the spring in the northern regions of Africa start to migrate northward across Europe. They do this as soon as they hatch from the chrysalis. Their journey can take as long as three months to complete, and during it they fly nonstop over the Mediterranean Sea.

However, even the painted lady migration is not the longest. This record belongs to the next butterfly we shall look at—the monarch.

The Monarch Butterfly

The monarch butterfly, which lives in the United States, Canada, Australia, and New Zealand, is a most amazing insect. Every fall, in

In the fall Canadian monarchs fly south. They are joined by others in the northern United States, so the migration becomes larger and larger. Those reaching Florida and southern California spend the winter in a less active state. The inland swarms fly on into Mexico.

Pacific Grove, California, is known as "Butterfly City, U.S.A." In the welcoming parade below, children have dressed as monarchs with wings and antennae.

August or September, millions of monarchs set off from Canada on an incredibly long journey. It is the longest known migration of any butterfly. Monarch butterflies migrate 1,900 miles from Canada to Mexico, where they stay for the winter. Then, in spring, they may travel 600 miles back to the north. A monarch butterfly born in Canada in August may live as long as 11 months and travel as far as 2,500 miles in that time.

This amazing feat has made the monarch butterfly famous. In some places in the United States the monarchs' migration is so well known that towns along the monarchs' route celebrate their passing with carnivals. One city, Pacific Grove, California, even calls itself "Butterfly City, U.S.A."

Why does the monarch make such a long journey? The butterfly migrates in order to avoid the cold weather. Once monarchs have reached their destination, they spend the winter among trees in evergreen forests. The sight of hundreds, often thousands, of monarchs clustered on the trees is spectacular. People call them "butterfly trees" and say the trees look as though they are "dripping" with butterflies.

On most days the butterflies stay on the trees and keep warm. They only leave the trees in order to find food, and always return at night to sleep.

In the fall, while migrating south, monarchs just feed, sleep, and migrate. They do not mate or lay eggs. In spring, as they migrate back to the north, they begin to lay eggs. Mating and egg-laying slows down the migrants. Most do not live long enough, or cannot fly fast enough, to get all the way back to their Canadian starting point.

However, the eggs laid in the

27

southern United States will turn into young butterflies that fly back to Canada in July or August. These butterflies then lay eggs in Canada. The young butterflies that hatch from these eggs begin another long

A butterfly tree. The same trees are used year after year. In some places the butterflies are protected by law from being disturbed.

journey to Mexico. So it takes three generations of butterflies to complete the migration cycle. It is the grandchildren of the butterflies in Mexico one winter that return to Mexico the next winter.

The Wanderers

In Australia the monarch is known as the wanderer. In the fall, just before the cold weather begins, wanderers begin to migrate. They move from the inland areas of Australia to the warmer coasts. They form "butterfly trees" in three areas—the Hunter Range northwest of Sydney, the Sydney Basin, and around Adelaide. In spring the second generation migrates back to the inland areas they came from.

Monarchs and wanderers can turn up in unexpected places. While on the Australian coast, some wanderers are lifted by the wind as high as four miles. Then they are blown across the sea to arrive in New Zealand. The North American monarchs can get blown so far off course that they turn up in Britain.

Dragonflies

Butterflies are not the only large insects that migrate long distances cross-country. Dragonflies lay their eggs in ponds and lakes. The dragonfly larva is called a nymph. It has no wings and lives underwater. The dragonfly nymph is a deadly predator, attacking and eating many other water insects. When it is one or two years old, the nymph crawls out of the water. Its skin splits open and the adult dragonfly, complete with wings, is born.

The adult dragonflies are also

A dragonfly usually flies a regular route near water, but it is powerful enough to go much farther. Some species cross the English Channel.

fearsome predators, chasing and eating many smaller insects. Like the butterflies we have already seen, many dragonflies migrate across the countryside. On their journey they visit one pond after another.

Finding the Way

Why is it that butterflies like the monarch hardly ever seem to get lost? How do they know which way to fly when they leave Canada? Butterflies and dragonflies find their way by using the sun as a guide. In the fall, monarch butterflies find their way south by flying toward the sun. The butterflies that make the return journey north fly in the opposite direction. You may have noticed that butterflies are hard to spot on a day when it is raining, or even when it is cloudy. This is because butterflies migrate only when the sun is shining.

How does such a small creature avoid being blown off course by the wind? The monarch is a very clever insect. When a strong wind threatens to blow the monarch in the wrong direction, it flies closer to the ground where the wind tends to be gentler. Sometimes it will fly along using woods or tall crops as shelter from the wind. However, if the wind is blowing strongly in the right direction, the monarch uses it to fly higher and faster. When the wind is just right, some monarchs can travel as far as 60 miles in a day.

5 Moths

Have you ever wondered what the difference is between a moth and a butterfly? They are very similar in appearance, and both have wings covered with thousands of tiny scales. These scales have a dust-like texture and rub off on your hand if you touch the wings. Both moths and butterflies feed on nectar found in flowers.

However, moths and butterflies do differ from each other in a number of ways. For example, most butterflies are brightly colored, but most moths are dull in color. Also, butterflies fly around during the day when the sun is shining, and sleep at night. Most moths, however, do the opposite.

They sleep during the day, and fly around only at night when it is dark.

Since moths fly at night, they are hard to see. This makes it very difficult to know what they are doing and where they are going. However, moths can be tracked by radar. Scientists can watch moths migrating on radar screens, and figure out where they are going.

It seems that the way in which moths migrate is very similar to the way butterflies migrate, except that

Many moth larvae, like butterflies, spin webs. Here a bush is almost covered with the larval webs of the small ermine moth.

moths fly at much greater heights than butterflies do. Butterflies migrate only a few yards above the ground, but moths may migrate hundreds of yards up in the sky.

Moths and Lights

You may have noticed moths flying into brightly-lit rooms through windows that have been left open. Perhaps moths have flown into your bedroom late at night. Moths are often attracted to the light this way. Once inside a room they fly around and around in a circle, gradually getting nearer to the light source. Often they end up bumping into the light bulb or settling in the folds of a lampshade. Moths are attracted to lights because of the way they find their direction when they are migrating.

We have already seen how butterflies find their way on their cross-country travels, by using the sun as a guide. Moths also need something to guide them. But because they migrate at night, they have to use the moon instead. Many moths fly cross-country always keeping the moon in front of them. This allows them to fly in a straight line and, of course, since the moon is thousands of miles away, they never bump into it. Sometimes, though, a moth gets confused and mistakes other lights, like your bedroom light, for the moon.

Many migrating moths are

Tricolor buck moths in Arizona cluster around a bright lamp, mistaking it for moonlight. Moths are attracted to light when migrating.

attracted to lights. So, next time a moth flies into your bedroom, just think: it might have flown 60 miles or more before flying in through the window.

Scientists have a clever way of catching migrating moths. They use a special kind of trap, called a

The beautiful yellow hind pair of wings can be seen partially concealed under the dull front wings.

"light" trap. Light traps confuse the migrating insect. A very powerful light bulb is placed above a funnel leading into a large box. Flat plates stick out from the side of the light. Moths that are attracted to the light bump into these plates and fall through the funnel into the box. There they are trapped until the scientist comes to look at them the next morning.

The Large Yellow Underwing Moth

The large yellow underwing moth is found all across the United States, Europe, and Asia. Its caterpillar feeds on all types of plants, and is sometimes regarded as a pest.

During the day the adult moth often sleeps in sheds and houses, sometimes behind curtains. It has dull, brown front wings. When it is sleeping, these dull, brown wings are all that you can see. However, if disturbed, the moth will jump and scamper or fly away with alarming speed. As it does so, it flashes the bright yellow underwings from which it takes its name.

Large yellow underwing moths

migrate cross-country from garden to garden, or meadow to meadow, just like the small white butterfly in Chapter 4. Yellow underwings are very strong fliers. During their three or four months of life they can travel as far as 175–250 miles. During heavy migrations as many as 3,000 large yellow underwing moths have been caught in a single light trap during just one night.

The Bogong Moth

The Bogong moth is found only in parts of Australia. In spring and fall the bogong lives on the Australian lowlands, mating and laying eggs on many types of plants. At the beginning of the hot summer, millions of bogong moths migrate across the lowlands to the Australian mountains. There they fly up into the mountains until they find cool caves in which to spend the long, hot summers. On a spring evening at the height of migration, thousands of bogong moths may be seen circling around outside each cave as the sun sets. In the fall they leave the caves, and migrate back down onto the lowland plains.

The bogong moth lives in southeastern Australia. In summer, millions swarm up into the mountains to find relief from the heat.

The Silver Y Moth

The silver Y moth gets its name from the silvery, "Y"-shaped mark on each of its pair of front wings. This type of moth is found all through Europe and Asia, and performs quite long migrations. Some silver Ys may travel as far as 600 miles.

In winter in Europe, the silver Y is found around the Mediterranean, in North Africa, Spain, and Italy. In spring, as soon as they emerge from the chrysalis, many silver Y moths begin to fly north. They cross France and Germany. Some moths cross the Baltic Sea and reach Norway and Sweden. Others cross the English Channel and arrive in Britain.

For the first five days of their life as a moth they fly during the day as well as at night. Then they settle down to a more leisurely pace. They fly only by night, mating and laying eggs as they go. The caterpillars of these northward-bound migrants feed on many types of plants, and turn into moths in the fall. These moths then head south and eventually wind up back in the Mediterranean region. There they lay eggs that produce moths the following spring. It is these moths that will once again head north.

The silver Y is not much to look at. It is only about an inch long and despite its silvery Y-mark, is not very pretty. However, some of the rarer moths that live around the Mediterranean and migrate to northern Europe look much more impressive. Some are also quite big, like the death's-head hawk moth, which is the size of a mouse and has a mark on its back that looks just like a human skull.

The humming bird hawk moth is also interesting to watch. Unlike most moths, it flies by day. It looks just like a tiny humming bird as it hovers in front of flowers, sipping nectar with its long tongue.

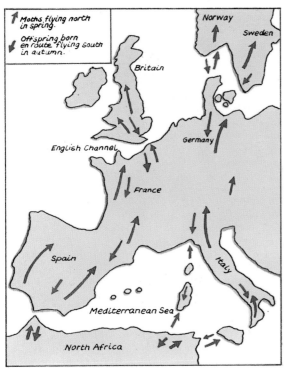

As soon as they emerge from the chrysalis in spring, some silver Y moths begin migrating north. Their caterpillars will become moths in the fall, and return to the Mediterranean.

How Moths Find Their Way

We have already looked at how moths use the moon as a guide to

The humming bird hawk moth looks like a tiny bird as it hovers to sip nectar. Hawk moths are probably the fastest insect flyers.

help them fly in a straight line when they migrate. Unfortunately for the moths, though, the moon does not shine every night. When the sky is cloudy the moon can disappear from view. Sometimes, even when the sky is not cloudy and the stars are shining, the moon still cannot be seen. On such nights, moths use the brightest stars as a guide instead of the moon. But what happens on very cloudy nights when the moths cannot see either the moon or the stars? They have to rely on another sense. Instead of using the moon and stars as a guide, they use their ability to sense "magnetism."

Our planet is like a huge, round magnet floating through space. It is surrounded by strange forces called "magnetic lines of force." It is these invisible magnetic lines of force that make a compass needle always point toward north. Moths, along with many other animals, have an amazing ability to sense and use magnetic lines of force to help them find their way.

6 Locusts

Locusts live in very hot and dry places around the world. Unless you happen to live in, for example, Africa, Australia, or South America, it is unlikely that you have ever seen a locust.

There used to be a type of locust living in North America but it has not been seen alive since the end of the nineteenth century. It was probably killed off by insecticides as people settled the country. However, the dead bodies of this type of locust can still be found. They have been "deep-frozen" in the ice in glaciers

Locusts swarm because they must find damp soil in which to lay their eggs. Here a female has pushed the egg-depositing organ (below her abdomen) four inches into the ground.

that are on the Rocky Mountains.

Although you may not have seen a locust, many of you have seen an insect that looks very similar. This insect is the grasshopper. In fact, a locust is a kind of grasshopper, although its behavior is much more spectacular.

Hoppers

Although locusts live in dry places, the female has to lay her eggs in wet soil. Just before egg-laying time the female's abdomen grows in length. To lay her eggs, she pushes her long abdomen deep down into the soil. She lays as many as 90 eggs at a

All locusts migrate, but not all do so in vast numbers. It is crowding that decides if they will become solitary locusts or not.

time. Each egg is about the size of a grain of rice. The female covers the eggs with a liquid. This liquid hardens around the eggs, and protects them. Usually, the eggs hatch within a week of being laid. However, they can survive unhatched in the soil for as long as three months and still produce healthy young locusts!

When it hatches, the young locust has to tunnel its way up through the soil and into the daylight. Young locusts are called hoppers. Hoppers are extremely greedy! They seem to spend all their time either eating or looking for more food to eat. In fact they grow so fast that they actually grow out of their skins. The old skin is shed and a new one replaces it. Before the hopper reaches adult size it has to shed its skin at least five times.

In some hoppers a very interesting change takes place before they become adults. If there are lots of hoppers living within a very small area, the food will soon run out. When this happens the hoppers set off in a large group, and hop across the land to find more food. These hoppers eventually grow into the "swarming" type of locust.

However, for those hoppers that grow up where there is plenty of food, life is much less complicated. They do not have to set off in search of more food. These hoppers grow into what are called "solitary" locusts. This type migrates on its own, whereas the swarming type moves around in a massive group called a "swarm." Also, the solitary locust is less brightly colored than the swarming type.

Solitary Locusts

Solitary locusts migrate very quietly and almost unnoticed. Just like moths, they fly high in the sky, and always at night. Also like moths, they spend much of the day resting and feeding.

The aim of every locust migration

A swarm of locusts can contain countless millions. If they leave any greenery behind, it will soon disappear when their hatchlings appear two or three weeks later.

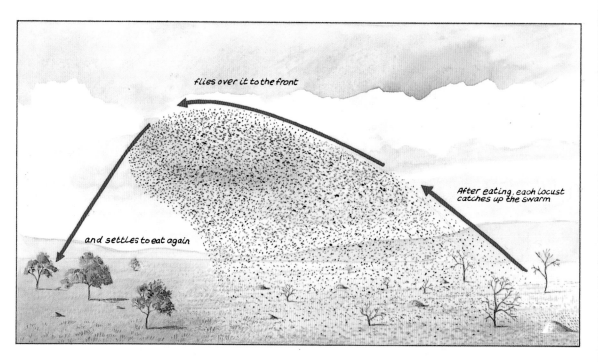

is to find wet soil. The locust will migrate each night until it finds an area in which rain has recently fallen. Here, after mating, the female lays her eggs in the soft, damp soil. Laying the eggs in such fertile soil ensures that, when the hoppers emerge, there will be plenty of young green plants for them to feed on. If there was only a small amount of old, dry vegetation most of the hoppers would almost certainly die of starvation.

Solitary locusts rarely live for very long. Most die after only a few weeks. During its short lifetime a solitary locust migrates a total of about 60 miles.

Swarming Locusts

The way swarming locusts migrate is spectacular indeed. Swarms are sometimes formed by other insects, such as bees and wasps, but they are never as large or as impressive as locust swarms. The largest swarms are in parts of Africa, where a single swarm may contain millions and millions of locusts.

Unlike the solitary locust, the swarming locust migrates during the day and is very easy to see. A locust

The only color left on this bush is that of the locusts themselves. A swarm can leave a trail of devastation like this for hundreds of miles.

Under a sky darkened by their numbers, a swarm of locusts brings fear of starvation to a Sudanese farm.

swarm is an awesome and even a frightening thing. It first appears like a cloud far away on the horizon. This is because the locusts at the top of the swarm are very high in the sky, sometimes as high as two miles above the ground. This towering wall of insects may be up to a half mile across.

The millions of locusts at the bottom of the swarm settle on the ground and eat every plant in sight.

tremendous—a sound like giant sheets of rustling plastic. A very big swarm may take as long as an hour or so to pass over. The last locusts to leave are those on the ground, having a final hurried bite to eat. As soon as they see the tail end of the swarm above them they take off into the air. These locusts have to chase after and catch up to the disappearing swarm, then overtake it before they can eat again.

What would it be like to be caught in a locust swarm, like the boys in the picture? The noise of so many large and powerful insects flying around would be deafening. Some would not be able to avoid bumping into you as they zoomed around. Since the locust is such a heavy insect this could hurt a little. However, locusts are not meat-eaters, and would not be at all interested in attacking you. The only concern of those near the ground is to land and eat as many plants as they possibly can. Even so, it would still be a very frightening experience.

Locust Devastation

After eating, they fly up to the top of the swarm again. The locusts at the top of the swarm are flying much faster than those farther down. The swarm is so large that, as the first locusts pass by, the sun is temporarily blacked out, turning the sky to a dark gray. The noise is

The millions of hungry locusts that settle on the ground ensure that every last scrap of greenery is eaten. The swarm is like a big vacuum cleaner, sucking up all the plants as it grinds along. Farmers in Africa and Asia, who rely on their crops to feed their families, are terrified that a locust swarm will visit their farm. If

41

it does, they and their families may starve. So, although locust swarms do not attack people directly, people may die of starvation after a locust swarm has passed through the area where they live.

The locust swarm travels for weeks until it finds rain. Since locusts live in such dry parts of the world, some swarms may need to migrate 2,000 miles or so before they find rain and eventually stop. Once on damp soil, the swarming locusts begin to mate and lay eggs, just like the solitary locusts. Hoppers soon appear, which rapidly become adults, and the cycle of migration and devastation starts all over again.

Because locusts can cause many people to starve, scientists have tried to find ways of controlling them. Since the 1950s the largest swarms have been tracked down by airplanes and sprayed with millions of tons of insect poison. So far, this seems to have worked. Very large swarms of locusts have not been seen in Africa since the 1960s.

In the Atlas Mountains of Morocco an airplane spreads pesticide over newly-sown fields. It is hoped that locust swarms are all a thing of the past.

Glossary

carnivore A carnivore is an animal that feeds mainly on the flesh of other animals.
chrysalis The pupa of a caterpillar.
egg The "shell" (which can be hard or soft, and is usually rounded or oval-shaped) from which young insects are born. The egg is produced from the mother's body and is placed on plants. When the insect is ready to be born it forces its way out.
evergreen forests Forests of trees or shrubs that keep their leaves all the year.
grub The legless larva of beetles, wasps, and some other insects.
home The habitat where an insect breeds is its *first* home. If it migrates seasonally, the place it goes to is the *second* home.
larva See **metamorphosis**.
mate/mating A female insect joining with a male insect to produce eggs. All insects are born as a result of their parents' mating.
metamorphosis Some creatures, including all insects, begin life in a completely different form from their parents. The change is called metamorphosis. It usually has four stages: 1. the female adult lays eggs; 2. the egg hatches into a larva that moves about and feeds; 3. the larva becomes inactive while it changes into a pupa; 4. the pupa splits open and a young adult comes out. The best-known metamorphosis is from butterfly's egg to caterpillar (larva) to chrysalis (pupa), to butterfly.
migrant/migrate/migration The habit of moving from one habitat to another (usually in search of food) is called *migration*. An insect that *migrates* is a *migrant*.
nectar A sweet liquid produced by some flowers, which insects drink.
north The direction a compass needle points. Most maps are drawn so that the northernmost part is at the top.
offspring The young insects produced by one adult female.
predator An animal that hunts and kills other animals for food.
pupa See **metamorphosis**.
south The direction opposite to north.
swarm A very large group of migrating insects, such as bees or locusts.
vegetarian A vegetarian is an animal that feeds only on plants.
wanderer The Australian name for the butterfly that in North America is called the monarch.

More Books to Read

Children's books containing some information on migration

Bender, Lionel *Spiders* Franklin Watts 1988
Blau, Melinda E. *Killer Bees* Raintree 1983
Dallinger, Jane *Grasshoppers* Lerner Publications 1981
—— *Spiders* Lerner Publications 1981
Dallinger, Jane & Overbeck, Cynthia *Swallowtail Butterflies* Lerner Publications 1982
Green, Carl R. & Sanford, William R. *Tarantulas* Crestwood House 1987
Hellman, Hal *Deadly Bugs and Killer Insects* M. Evans & Co. 1978
Insect World Silver Burdett Press 1989
Kerby, Mona *Friendly Bees, Ferocious Bees* Franklin Watts 1987

Lisker, Tom *Terror in the Tropics: The Army Ants* Raintree Publishers 1983
Mitchell, Robert & Zim, Herbert S. *Butterflies and Moths* Western Publishers 1987
O'Toole, Christopher *Discovering Ants* Franklin Watts 1986
Overbeck, Cynthia *Ants* Lerner Publications 1982
Patent, Dorothy H. *Mosquitoes* Holiday House 1986
Porter, Keith *Discovering Butterflies and Moths* Franklin Watts 1986
Robson, Denny *A Closer Look at Butterflies* Franklin Watts 1986
Whalley, Paul *Butterfly and Moth* Knopf 1988

Picture Sources

ARDEA: 9; Ian Beames 17, 24, 32, 35; Jean-Paul Ferrero 33; Bob Gibbons 20; J-M Labat 37; Ake Lindau 23; B. Sage 13.

Bruce Coleman: Jen and Des Bartlett 31; Jane Burton 12; A.J. Deane 15; M.P.L. Fogden 16; Jeff Foott cover and 6-7; D. Houston 36; Frans Lanting 26-27, 28; Prato 11; Hans Reinhard 21; Konrad Wothe 29.

Oxfam: 39; Peter Strachan 40-41

Simon Girling Associates/Richard Hull: 6, 8, 14, 18, 19, 22, 26 upper, 34, 38.

Wildlife Matters 25.

ZEFA: W.H. Davidson 30; Tortoli 42.

Index

A
Adelaide, Australia 28
adult stage of metamorphosis 8
Africa 14, 25, 36, 39, 42, *see also* North Africa
ant, *see* army ant
aphid 17-19
army ant 14-16
Asia 32, 34
Australia 25, 28, 33, 36

B
Baltic Sea 34
bee 10-11
beet leafhopper, *see* leafhopper
bogong moth 33
Britain 28, 34
butterflies 21
   monarch 8, 25-28, 29
   painted lady 25
   small white 23-24, 25
   wanderer 28
Butterfly City, U.S.A. 27

C
cabbage butterfly, *see* small white
California 27
Canada 25-28
caterpillar 9, 10, 14, 20, 23, 32
chrysalis 24, 34, 43, *see also* pupa
colony
   army ant 14-16
   bee 10-11
   wasp 11-12

D
death's head hawk moth 34
dragonfly 8, 28-29

driver ant, *see* army ant

E
eggs and egg-laying 7-8, 43
   butterfly 22, 23, 27-28
   dragonfly 28
   hunting wasp 12
   leafhopper 19
   locust 37, 39
   moth 20, 33
English Channel 34
Europe 25, 32, 34

F
food storage
   by bees 10
   by wasps 11-12
France 34

G
Germany 34
grasshopper 37
grub 43, *see also* larva

H
homes 9-12, 13-15, 43
   army ant 14-16
   caterpillar 10
   honeybee hive 10
   wasp 11-12
honeybee 10-11
hopper, *see* leafhopper
hopper (young locust) 37-38, 42
humming bird hawk moth 34
Hunter Range, Australia 28
hunting wasp 10, 11-12

I
insect, definition of 7
Italy 34

L
ladybug 6
large yellow underwing moth 32-33
larva 8, 43, *see also* grub
   bee 10
   dragonfly 8, 28
   leafhopper 19
   wasp 12
leafhopper 19-20
light trap 31-32
locust 8, 36-42
   hopper 37-38, 42
   solitary locust 38-39
   swarming locust 38, 39-42

M
magnetic lines of force 35
mating 22, 27, 33, 34, 43
Mediterranean Sea 25, 34
metamorphosis 7-8, 43
Mexico 26-28
migration
   alone 8, 21
   by night 8, 34
   definition of 2, 8, 43
   drifting 17-20
   flying 8, 18, 22, 24-28, 33
   hopping 19
   in swarms or other large groups 8, 14, 16, 39-42
   scent trail 15-16
   walking 8, 14, 16, 23-24
monarch butterfly 8, 25-28, 29
moths 20, 30-35
   bogong 33
   death's head hawk 34
   difference from butterflies 29-30
   humming bird hawk 34, 35

large yellow underwing 32-33
silver Y 34

N
navigation
   butterfly 29
   moth 30, 34-35
   wasp 12
nectar 10, 22, 43
New Zealand 25, 28
night flying 8
North Africa 25, 34
North America 36, *see also* United States and Canada
Norway 34
nymph of dragonfly 28

O
oceans and seas
   Baltic 34
   Mediterranean 25, 34

P
Pacific Grove, California 27
painted lady butterfly 25
predators of insects 43
   army ant 16
   dragonfly 28-29
   hunting wasp 12
   swallow 18
pupa 8, 24, 43, *see also* chrysalis

Q
queen ant 15

R
Rocky Mountains 37

S
scent trail 15
seas, *see* oceans and seas
silver Y moth 34
small white butterfly 22-24, 25

soldier ant 14-16
South America 14, 36
Spain 34
swarm 8, 43
Sweden 34
Sydney Basin, Australia 28

T
tent home 10
Texas 19
thistle butterfly, *see* painted lady butterfly

U
United States 19, 25-28, 32, *see also* North America

W
wanderer butterfly 28, 43
wasp 10, 11-12
webs, 10, 14
wild bee 10
worker ant 15

© Copyright 1991 Young Library Ltd.
Corsham, Wiltshire, England

MAY 1996

```
J 595.7
ORA
Oram, Liz.

Insect migration /
Steck-Vaughn.   c1991.
```

3 9110 00076407 8
GLEN ROCK PUBLIC LIBRARY, Glen Rock, N.J.
a39110000764078b

FINES WILL BE CHARGED IF CARDS
ARE REMOVED FROM POCKET

Glen Rock Public Library
315 Rock Road
Glen Rock, N.J. 07452

201-670-3970